Examining Geothermal Energy

Jordan Boyle

First published in 2013 by Clara House Books, an imprint of The Oliver Press, Inc.

Clara House Books
5707 West 36th Street
Minneapolis, MN 55416
USA

Produced by Red Line Editorial

The publisher would like to thank Cathy Chickering, SMU Geothermal Laboratory Project Specialist, for serving as a content consultant for this book.

Picture Credits

Fotolia, cover, 1; Joe Gough/Fotolia, 5; Shutterstock Images, 8, 10, 14, 18, 20, 25, 29, 32, 41; Tamara Kulikova/Shutterstock Images, 12; Anita Potter/Shutterstock Images, 13; Lukáš Hejtman/Shutterstock Images, 17; Tomasz Parys/Fotolia, 22; iStockphoto, 23; Galyna Andrushko/Shutterstock Images, 26; Christine Langer-Pueschel/Shutterstock Images, 28; Jonathan Hill/iStockphoto, 30; Ryan Brennecke/The Bulletin/AP Images, 34; Jim Parkin/Shutterstock Images, 36–37; Steve Bower/Shutterstock Images, 39; Rob Byron/Shutterstock Images, 45

Library of Congress Cataloging-in-Publication Data

Boyle, Jordan.
Examining geothermal energy / Jordan Boyle.
pages cm. -- (Examining energy)
Audience: Grades 7 to 8.
Includes bibliographical references and index.
ISBN 978-1-934545-41-6 (alk. paper)
1. Geothermal engineering--Juvenile literature. I. Title.
TJ280.7.B69 2013
621.44--dc23
2012035244

Printed in the United States of America
CGI012013

www.oliverpress.com

Contents

CHAPTER 1

Earth Energy

It takes a great deal of energy to heat homes, light buildings, and power vehicles. Today, more than 80 percent of the world's energy comes from non-renewable sources such as fossil fuels. Coal, oil, and natural gas are all fossil fuels. They are made from organic material that has been buried underground for hundreds of millions of years. The earth has a limited supply of fossil fuels. Using these sources of energy can harm the environment.

Scientists and researchers around the world are constantly looking for better ways to use energy. They are seeking new energy sources, as well as ways to use current energy sources more efficiently. These alternative energy innovators are working to balance energy consumption against the needs

Geothermal energy harnesses the heat of the earth and uses it to produce energy.

of our environment. Many innovators believe using more geothermal energy could help us achieve this goal.

Geothermal energy, which means energy from the heat of the earth, is a renewable energy resource. The earth stores heat from the sun. Just below its surface, the earth maintains a stable temperature of about 50 degrees Fahrenheit (10°C). The earth produces heat from within. In some places, liquid rock, called magma, flows near the surface of the earth, heating up the rocks that make up the earth's crust. Geothermal energy uses this heat from within the earth for heating and cooling, as well as to make electricity. Many homes and businesses across the United States and the world already use geothermal energy. Many people believe we should be using geothermal energy even more.

EXPLORING GEOTHERMAL ENERGY

In this book, it is your job to learn about geothermal energy and its role in our energy future. How long have humans been using geothermal energy? How can heat from within our planet power a computer or heat your house? Are there different ways to use geothermal energy? How might we use it in the future?

Derek Park wants to spend his winter vacation learning more about geothermal energy. He will travel around the world, meeting with scientists and energy experts. Reading Derek's journal will help you in your own research.

CHAPTER 2

How Heat Moves

My first stop is the science lab at Union Community College in Wisconsin. I'm meeting with physics instructor Jason Stevenson to learn more about geothermal energy.

"Good afternoon, Derek," Mr. Stevenson says as he leads me to a lab bench where he's set up a series of experiments. "The word *geothermal* means 'Earth heat.' So in order to understand geothermal energy, you need to know something about how heat energy behaves."

A teakettle is whistling over a small Bunsen burner. Mr. Stevenson turns off the burner and pours the water into an insulated mug. He hands me a thermometer. "How hot is the water?"

I measure it. "The water is 212 degrees Fahrenheit."

"Or 100 degrees Celsius—very good, Derek," Mr. Stevenson opens his freezer and takes out a metal spoon. "What temperature is the spoon?" he asks.

The earth's interior is very hot. Sometimes this heat finds its way to the surface, where it can be used to produce power.

I touch the thermometer to the spoon. "It's 32 degrees Fahrenheit, or 0 degrees Celsius."

"So the water is much hotter than the spoon," Mr. Stevenson says. "Now watch what happens when I put the spoon into the hot water."

We wait five minutes. Then I measure the temperature of the spoon and the water again. The water is cooler, and the spoon is warmer.

"Now do you understand how heat moves?" Mr. Stevenson asks.

"It moves from hotter things to colder things," I tell him.

"Right-o! Now let's try another experiment." Mr. Stevenson relights the Bunsen burner. Before long, the teakettle is whistling. He holds a metal pinwheel near the steam pouring out of the teakettle's spout. The pinwheel begins to spin. "Heat is actually a form of energy. Right now, I'm using heat energy to make the pinwheel spin," Mr. Stevenson says.

He tells me that energy cannot be destroyed—it can only change from one form to another. The heat energy from the burner boiled the water. The energy was turned into motion when the water formed steam, a gas. Motion is a form of mechanical

FORMS OF ENERGY

There are many different forms of energy. Mechanical energy has to do with the motion of objects. Nuclear energy is the energy from radioactive particles. Electrical energy is the energy related to electric charges. Heat is thermal energy. Cold is simply a lack of heat energy. All energy is either kinetic or potential. Something's stored energy is its potential energy. Kinetic energy happens when something is actually in motion.

The kinetic energy of moving steam can be turned into mechanical energy. Energy can never be lost, only turned into a new kind of energy.

energy. The mechanical energy from the moving steam powered the pinwheel blades. Steam is an important part of geothermal energy.

Mr. Stevenson tells me to keep these experiments in mind as I investigate ways of using geothermal energy. I thank him and say goodbye. I'm flying off to Pompeii, Italy, to visit the Roman baths.

CHAPTER 3

Ancient Geothermal Energy

Mount Vesuvius towers above me. I'm on a group tour of the Stabian Thermae ruins, the oldest public baths in Pompeii. They were built in the fourth century BCE.

Our guide, Gina Napolitano, greets us. "Baths were an important part of Roman culture. The number of hot springs in this area made it an attractive place for the Romans to build a city."

She takes us inside the men's caldarium, or hot bath room. "Romans were famous for their public baths. They built them across Europe. Most were heated by burning wood furnaces. But here in Pompeii, the Romans didn't have to burn wood. They built these public baths around natural hot springs. Geothermal energy heated the water."

Ancient Romans built public baths across their empire. Some of these were warmed using geothermal heat.

We leave the public baths and head down the cobblestone street, walking past the ruins of the ancient city.

"The buildings of Pompeii were heated by hot springs as well," Ms. Napolitano says.

"What creates hot springs?" I ask.

"Hot, liquid rock, or molten magma, from within the earth flows through cracks in rock near the earth's surface. The magma heats water trapped in underground reservoirs, creating hot springs."

"Were the Romans the only ancient people who used geothermal energy?" I ask.

Ms. Napolitano shakes her head. "Not at all. American Indians used geothermal energy more than 10,000 years ago. They bathed in hot springs, and they cooked their food in hot springs, too."

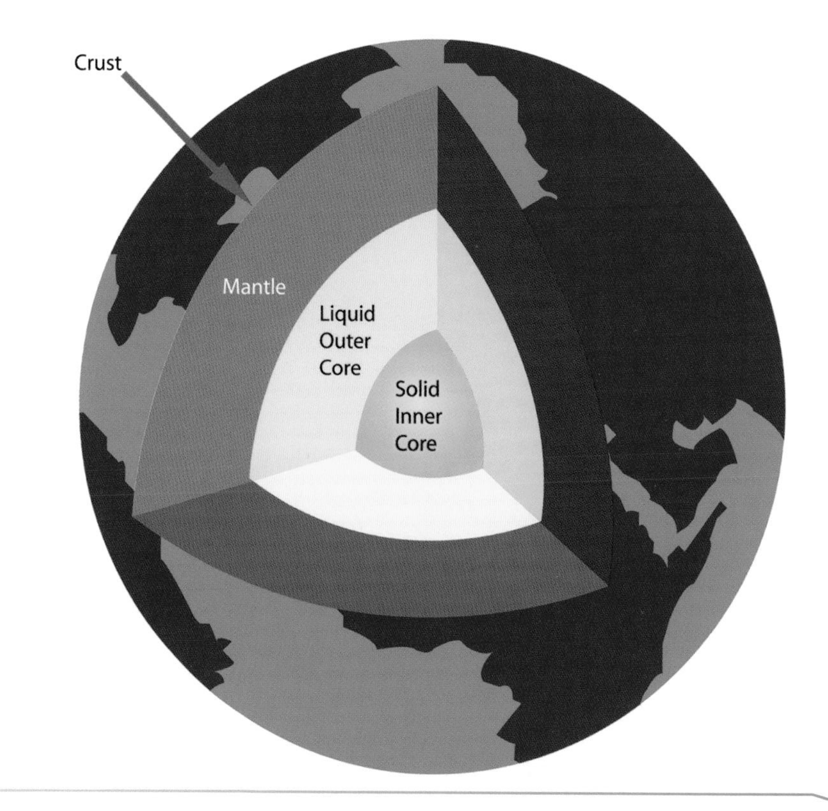

"Are there hot springs all over the world?" I ask.

Ms. Napolitano tells me hot springs are pretty rare. They usually only occur at active geological

THE EARTH'S CORE

Several different processes produce heat deep in the center of the earth. Some heat is produced in the core. Earth's core has two parts. One is solid. The other is melted rock called magma. Next to the core is the mantle. The mantle is also full of magma. Most of the earth's heat is stored in the mantle. Usually, the magma stays below the earth's crust, where it heats surrounding rocks. Sometimes, the magma breaks through to the surface, forming volcanoes. The heat mainly comes from radioactive decay. It flows into the next layer of the earth, the crust, where we can access it.

sites. These sites are mostly found along the edges of tectonic plates and at thin hot spots in the earth's crust. Tectonic plates are pieces of the earth's crust.

I had no idea the earth was such a fiery place. I bid Ms. Napolitano goodbye. To learn about how geothermal energy is used today, I am headed to another volcanically active site.

THE RING OF FIRE

The Ring of Fire earned its name from its many volcanoes. About 75 percent of the earth's 1,500 volcanoes are located in this region. The Ring of Fire is defined by tectonic plates under the Pacific Ocean pushing against continental plates. The region runs down the west coast of Canada, the United States, and South America, and along the east coast of New Zealand, Indonesia, the Philippines, and Japan. Earthquakes are more common in the Ring of Fire than other places on the earth.

Sometimes magma forces its way to the earth's surface when volcanoes erupt.

CHAPTER 4

Heat from Within

Today I am meeting with Heddy Palsdottir, who works for the Iceland Energy Agency. She is showing me a pool of water called the Blue Lagoon, which is located outside of Reykjavik, Iceland. I follow her across a wooden bridge to a platform, where we slip into the Lagoon's deep blue waters.

It's a cold day, so I expect the water to be freezing. Instead, it feels like I'm sitting in a hot tub. "What's the temperature of this water?" I ask Ms. Palsdottir.

"About 102 degrees Fahrenheit, or 39 degrees Celsius," she says. "Iceland sits on top of the Mid-Atlantic ridge. This is a place where plates under the Atlantic Ocean are pulling apart. Magma is constantly seeping up from the spreading crack. When the magma reaches the surface, it cools and becomes solid rock. This cooled magma actually formed the island of Iceland

Iceland's Blue Lagoon is famous for its Earth-heated warm waters.

millions of years ago. Iceland is a very active geologic area. We have about 35 volcanoes and 600 hot springs."

"So the Blue Lagoon is a natural hot spring?" I ask.

Iceland was formed by magma seeping up over millions of years.

"No, it's man-made," Ms. Palsdottir points to a power plant at the far end of the lagoon. "This water is pumped from a well more than a mile below the earth's surface. The well is deep enough to be heated by geothermal energy. We use the hot water from the well to heat nearby homes and generate electricity. Then it's released into the lagoon."

Ms. Palsdottir tells me most Icelandic homes don't have furnaces. Instead, 89 percent of Icelanders use piped geothermal water to heat their houses. This is called direct use of geothermal energy because the heated water is coming directly from inside the earth.

I want to know if there are any problems with direct use of geothermal energy. I'm surprised when Ms. Palsdottir tells me not all geothermally heated water is safe for people and animals to use. Water heated deep in the earth dissolves gases and minerals from the rocks it passes through. Depending on what's in those rocks, the minerals and gases can sometimes be harmful to plants and animals. Pollutants must be filtered out of geothermal wastewater before it's released into the environment.

I know that alternative energy research often involves finding ways to replace fossil fuels with other energy sources. "Has using geothermal energy cut Iceland's use of fossil fuels?" I ask Ms. Palsdottir.

BOISE, IDAHO

Boise, Idaho, is a pioneer in the direct use of geothermal energy. In 1892, the citizens of Boise started piping water from a nearby hot spring to heat some of their homes and businesses. Now they have geothermal systems heating homes, businesses, and government offices throughout the Boise area.

"Yes," she says. "Iceland is the world leader in using renewable energy. In fact, we generate 100 percent of our electricity from renewable sources. More than 20 percent comes from geothermal energy. And you can see the difference." She points to the sky. "The area around Reykjavik

The renewable energy sources of Iceland produce less pollution than non-renewable sources such as fossil fuels, leading to clear skies.

was once very smoky because the air was so polluted. Now the air is clear."

Wow! It seems like geothermal energy has many possibilities. It's time to head back to the United States to learn more about geothermal power plants.

CHAPTER 5

Fountains of Steam

Today I'm at one of the many geothermal power plants just north of San Francisco, California, in an area known as the Geysers. Electrical engineer Juan Pasos has offered to show me around the station.

Mr. Pasos sweeps his arm across the horizon. "This area has high geothermal activity. Magma flows close to the earth's surface here. It heats underground water trapped within the earth's crust. The crust acts like the lid on a pot, keeping the hot water under pressure. When released through a hole to the surface, the hot water shoots out as steam."

He points to the white pipes snaking out across the ridgeline. "We drill beneath the crust and suck the earth-heated steam up through pipes and into our plant."

We follow the pipes as Mr. Pasos leads me into the building. "The steam passes through a separator to remove dirt and rock

Geysers are places where heated underground water breaks through to the surface as a jet of steam.

particles. Then it flows into power generators, where the steam spins the blades on a turbine. The turbine's mechanical energy forces the movement of electrons. This creates electricity."

I remember the pinwheel in Mr. Stevenson's laboratory. Steam is kinetic energy that can be transferred to create motion in another object.

We walk to another section of the building. "After the steam is used to spin the turbines, it is captured and cooled in huge towers," Mr. Pasos explains. "Some of the water evaporates, but the rest is recycled. We send the water back through some more pipes into the earth to refill the underground reservoir. There, the earth will reheat the water so we can reuse it to produce more power."

Geothermal power plants are able to recycle the water they use.

"So geothermal power is a renewable resource?" I ask.

"If you use it carefully," Mr. Pasos says. "The first commercial power plant was built here in 1960. Back then, we released our used geothermal water on the surface after making electricity. But by the end of the 1980s, our power production dropped. Our source of underground hot water was drying up."

"Is that why you put the water back underground now?" I ask.

"Yes, and it helped replenish our steam source," Mr. Pasos points to a pipeline leading over the horizon. "But some geothermal water is lost to evaporation. We had to find a way to replace that water. We did it by linking with local communities. They had wastewater, water from sewers or that had been used in manufacturing. They needed to get rid of this wastewater safely. So we came up with a way to treat the wastewater and pipe it back underground to refill our geothermal reservoirs."

Mr. Pasos explains that keeping the geothermal water away from the surface helps in another way, too. Water that flows through rocks deep underground can dissolve toxic chemicals and gases. I remember Ms. Palsdottir in Iceland explaining how this can lead to pollution. By putting the water back underground, in what is called a closed-loop system, we reduce the potential for polluting streams, lakes, and rivers with toxic chemicals.

LARDERELLO, ITALY

The first time geothermal energy was converted into electricity was in 1904 in Larderello, Italy. By 1913, a commercial power plant was operational. The area has been active since ancient Romans lived there. Steam from hot liquids underground was harnessed to create electricity. However, many geologically active areas cool over time. Larderello's hot spots have cooled over the years. In 2003, the plant realized its underground liquid reserves had been reduced by about 30 percent since the 1950s. The power plant is still in operation today, about 100 years after being built, providing electricity for approximately one million homes.

"How does geothermal energy compare to other renewable energy sources, like solar and wind power?" I ask.

"Geothermal power is more dependable," Mr. Pasos says. "Solar cells can only generate power when the sun shines. Wind turbines only produce power when wind blows. Geothermal power is generated around the clock."

The power plant in Larderello, Italy, was the first in the world to make electricity from geothermal energy. Humans have used the hot underground water in the area since ancient times.

He continues, "A geothermal energy plant also has a smaller footprint, which means it uses less land and resources than wind or solar farms."

Geothermal energy sounds great. I ask Mr. Pasos if there are any downsides.

"Well, no energy source is perfect. Once you get the plant up and running, it costs about the same as a coal-powered plant to operate, but the costs to build a geothermal plant are quite high," Mr. Pasos points to a drill in the distance. "We have to dig deep into the earth's crust. Sometimes we drill a hole and

Geothermal plants take up less space than other renewable-energy power plants. The earth's heat is a constant source of energy.

don't find enough hot water to power a plant. Then we have to try drilling someplace else. That's expensive. Of course, the high start-up costs are offset by the fact that you don't have to buy any fuel, like coal, oil, or gas, to run your power plant. The earth's heat provides the fuel."

"So why don't we use more geothermal power?" I ask.

"With our current technology, practical sites for geothermal electricity production are limited," Mr. Pasos explains.

"Can we use geothermal energy for other things besides generating electricity?"

"Oh, yes!" Mr. Pasos says with a smile. "You can use geothermal energy to power heat pumps just about anyplace in the world."

CHAPTER 6

Staying Warm with Geothermal Power

Today I'm visiting Memorial Middle School in Omaha, Nebraska. Patricia Hilstrom, the school's principal, stands to greet me as I walk into her office. "I take it you're here to look at our geothermal heat pump system. Let's have a look!"

As we walk, I ask Ms. Hilstrom, "Is Omaha near an active geothermal area?"

"You mean like a fault line between tectonic plates or a volcanic hot spot where magma wells up from Earth's core?"

I nod.

Ms. Hilstrom laughs. "Oh no, not at all. But that's the beauty of our geothermal system. We use the heat stored below the frost line. Everything above the frost line freezes in the winter. But the ground below the frost line stays warm.

Geothermal heating pipes are laid underground near the buildings they will heat.

You don't have to be near an active geothermal area to use this heat. The heat pump uses the constant temperature of the earth to heat the school."

She leads me to a basketball court. "We're standing on top of the first part of our geothermal heat pump—the heating coils," Ms. Hilstrom says. "The coils are buried under our basketball court, below the frost line. In this part of the country, the frost line is about three feet, or one meter, deep. Come inside, and I'll show you the rest of our unit and explain how it works."

We head to the school's utility room, where Ms. Hilstrom pats the side of a metal box. It has pipes flowing into it on the bottom and it's connected to ductwork at the top. "This ground source heat pump, known as a GSHP, heats our school and provides our hot water, too."

The GSHP is big, but the room is completely silent. I guess geothermal energy is quiet. Ms. Hilstrom says this is one of the benefits of a GSHP. The pumps make hardly any noise.

"The earth is great at storing that energy," she says. "Just below the frost line, the earth maintains a fairly constant temperature. Around here, it's an average of 50 degrees Fahrenheit, or 10 degrees Celsius, year-round. Our GSHP allows us to use that energy."

Ms. Hilstrom takes a deep breath. "GSHP systems aren't all good. Our system cost a lot of money to install. But there are no fuel costs. We never have to buy oil or natural gas. The earth provides our energy. Over time, this system actually ended up saving us money."

She tells me that while geothermal energy is pretty environmentally friendly, the unit does use some electric fans. Coal-powered plants generate the electricity the GSHP system uses. So it still emits some carbon dioxide. But it

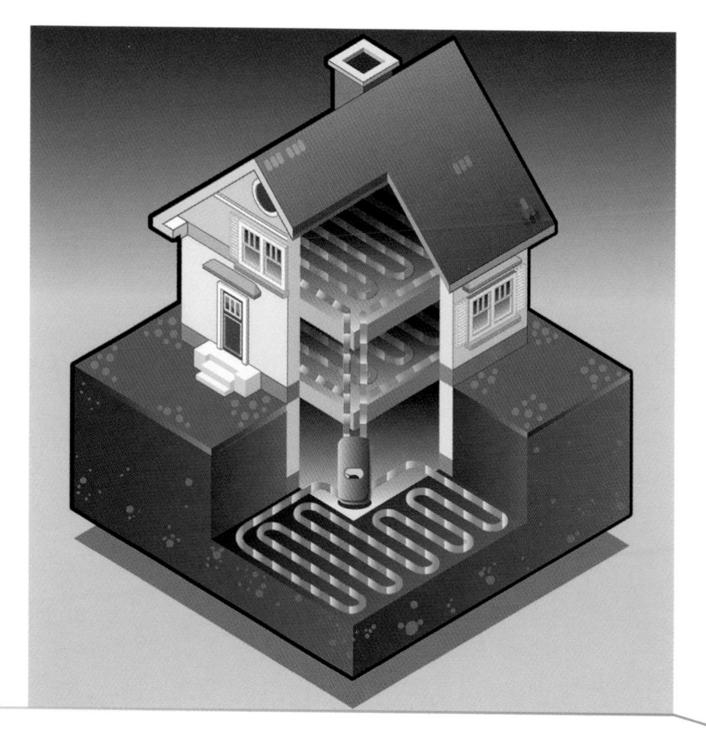

GROUND SOURCE HEAT PUMPS

A ground source heat pump (GSHP) uses a closed-loop system. The loop contains a liquid, which passes through coils under the frost line. The earth warms the liquid. Eventually, the liquid moves into the building, where it can heat the home and the water used for cooking and bathing.

Geothermal heat is a year-round energy source.

doesn't use nearly as much electricity as the school's old furnace, air-conditioner, and hot water heater used.

Ms. Hilstrom adds, "In summer, they can flip a switch to reverse the heat pump. The GSHP keeps the school cool, too!"

I thank Ms. Hilstrom for her time. My next stop is France, where I've heard scientists are exploring new ways to use geothermal energy.

CHAPTER 7

Geothermal Innovations

I'm at a drilling site in the rolling hills of France. A white and red drilling rig towers above me. Geophysicist Carla Minot is my guide today.

"Welcome to our project researching enhanced geothermal systems, or EGS." She hands me a hard hat.

"Thank you!" I say as I put it on and fasten the chin strap. "First, can you tell me what an enhanced geothermal system is?"

"Certainly. Geothermal energy is everywhere under our feet. But with current technologies, it's only practical to use it to generate electricity where we find three things." Ms. Minot ticks them off on her fingers: "One: High crustal temperatures. Two: An underground source of water close to the magma. Three: Rock permeable enough to allow water to flow through it. That combination of factors, which we geophysicists call conventional hydrothermal resources, only occurs in a few places on the earth."

Magma is molten, or hot liquid, rock. Once it reaches the earth's surface, it is known as lava.

"Like at the Geysers in California?" I ask.

Ms. Minot nods. "The United States is the world's leader in producing electricity through geothermal energy. Yet your country gets less than 1 percent of its electricity that way. Geothermal power has great potential. It's clean, renewable, and dependable. But we need to come up with technology that makes its use practical in more parts of the world."

She points to the ground. "Here in France we have rock heated by magma, but we don't have an underground water source or permeable rock. So we are experimenting with ways to make our own—using EGS."

Ms. Minot says the first step is to drill a hole into the hot dry rock. Then, her colleagues pump fluid into the hole, which fractures, or breaks up, the rock. This makes the rock permeable. Water can be pumped into the fractured rock, creating an underground reservoir of hot water.

"So what are you working on now?" I ask.

"Here we have high-temperature rock at fairly shallow depths. We believe we'll be able to enlarge natural faults already in the rock to create our artificial geothermal reservoir. We call it 'enhanced' because we're taking what exists in nature and changing it to create a geothermal system where one didn't exist naturally." She points to the drill rig. "The plan is to have one injection well where we pump in water and other wells where we withdraw heated water. Eventually, this hot water could be used to spin turbines and generate electrical power."

Ms. Minot says her team has to proceed

EARTHQUAKE DANGER?

Shallow geothermal drilling usually doesn't cause problems, but many scientists want to drill deeper wells to reach geothermal reservoirs. These deeper wells can create a higher earthquake risk for the surrounding area. In 2009, a geothermal plant in Basel, Switzerland, was shut down after a series of small earthquakes. Citizens concerned about earthquakes protested a deep drilling project near the Geysers, California. The project was eventually shut down. Most drilling-related earthquakes occur during the initial fracturing of rock rather than during the ongoing operation of a power plant. These earthquakes are usually minor enough that people on the surface rarely feel them.

Workers drill a test well at a possible enhanced geothermal well site in Oregon. Researchers hope to access geothermal energy at this site by using water to create artificial fractures in the rock.

carefully. Rocks deep below the earth's surface don't always behave in predictable ways. Injecting water into fault lines to make the rock permeable can trigger earthquakes. Some similar projects have been stopped due to public concern for earthquake risk.

Ms. Minot smiles. "But we are making progress. Many countries have come together to work on this process. The United States, Japan, the United Kingdom, Germany, and Australia have all conducted experiments in EGS. The more research we do, the closer we get to increasing our geothermal energy sources."

I thank Ms. Minot for her time. I'm off to an oil field testing center in Wyoming.

CHAPTER 8

The Future of Geothermal Energy

My last stop is the Rocky Mountain Oilfield Testing Center (RMOTC) in Wyoming. I'm talking with Dr. Nate Lodge from the U.S. Department of Energy. The hills around us are dotted with oil rigs.

"Welcome!" Dr. Lodge begins. "Here at RMOTC, we're exploring ways to use the water by-product of oil wells to produce geothermal energy."

"What does 'water by-product' mean?" I ask.

"We estimate that for every barrel of oil produced by an oil well, ten barrels of hot water are produced. Right now, oil companies see that water as a waste product. They have to spend money to get rid of it. If we could find a way to turn that water into something useful, it would be great."

“So you might be able to turn that water into electricity?” I ask.

“That’s right. Commercial geothermal power plants already use very hot water or steam from deep within the earth. But the water from our oil wells isn’t quite hot enough to make steam and spin a turbine to power a generator. So we’re looking for ways to generate electricity using cooler water. When oil comes up from a well, it usually mixes with water and other fluids. Here, we use the warm water that comes up with the oil to heat

Right now, fossil fuels are our most common energy source. The hot water produced from oil wells also could be used as an energy source.

a secondary fluid that boils at a lower temperature than water. Steam from that fluid turns the blades of a turbine to power our generator."

"So have you been able to use this method to produce electricity?" I ask.

"Indeed we have. And now commercial oil companies are slowly starting to use the technology we developed here. Some people call this co-production because we're producing both oil and electricity at the same site."

"Do your geothermal units produce a lot of electricity?" I ask.

"Not compared to a commercial geothermal plant," Dr. Lodge tells me. "But one of the problems of geothermal energy is that it's expensive to drill deep enough to find a geothermal resource. The holes drilled for oil wells are already here, so we eliminate much of the cost. And there are thousands of aging oil wells all over the country. If we could use them to generate electricity, it would be a big advantage."

"Are there other experimental geothermal technologies?" I ask.

"Oh yes! It's a very exciting field," says Dr. Lodge. "A company in the Southwest is experimenting with using carbon dioxide as a geothermal fluid instead of water. In many ways, carbon dioxide actually works better than water. It stores more heat, and less energy is needed to pump it."

GEOTHERMAL ENERGY AROUND THE WORLD

In 1988, the International Geothermal Association was established to encourage research and development of geothermal energy around the world. The association is exploring ways to adapt oil-drilling technology to geothermal production, identify new sources of geothermal energy, and improve current geothermal energy production. Geothermal energy use is expanding in many countries, including China, Japan, Indonesia, Kenya, El Salvador, Honduras, and Mexico.

In dry places, such as the Arizona desert, water can be hard to come by. Scientists are experimenting with using carbon dioxide to store heat instead of water.

Dr. Lodge goes on to say that water is scarce in the deserts of Arizona, but carbon dioxide is plentiful—too plentiful, in fact. It's one of the pollutants emitted by burning fossil fuels. Many scientists believe carbon dioxide is leading to the gradual warming of the earth's climate. If we could find a way to reduce some of the excess carbon dioxide by pumping it into the earth, we could help our environment while creating electricity at the same time.

"Sounds like there are a lot of exciting new developments on the horizon for geothermal energy," I say.

"Yes, there are!" Dr. Lodge shakes my hand. "Thanks for stopping by. I enjoyed discussing the potential of geothermal energy."

CHAPTER 9

Your Turn

You've had a chance to follow Derek as he did his research. Now it's time to think about what you learned. You discovered that since ancient times, people have used geothermal energy for bathing, heating buildings, and cooking. You learned that we still use geothermal energy for direct heating, but also to generate electricity.

You found out that most geothermal electricity production is currently in geologically active areas. But the geothermal energy just below the frost line is available everywhere. This type of geothermal energy can be used to heat buildings through ground source heat pumps. You also learned about experimental techniques for expanding the use of geothermal energy. While research has a long way to go, scientists are working hard to give geothermal energy a larger role in our energy consumption.

What do you think? Is geothermal power a practical alternative energy source?

YOU DECIDE

1. Why is geothermal energy more dependable than other alternative energy sources?
2. What are some of the downsides of geothermal power? Do you think geothermal power is worth pursuing anyway? Why or why not?
3. Which ways of using geothermal power are feasible in your area: direct heating, electrical generation, or ground source heat pumps? Why?
4. Do you think geothermal power will play a large role in our energy future? Why or why not?
5. How could you change your own behavior to help cut down on your energy use?

GLOSSARY

caldarium: An ancient Roman hot bath.

core: The central, innermost section of the earth.

crust: The outer layer of the earth.

electron: A negatively charged particle that can create electricity.

frost line: The ground depth below which frost doesn't go, keeping the ground at this depth a constant temperature.

generator: A machine that converts mechanical energy into electricity.

geothermal: Heat generated by the earth.

heat pump: A machine that moves heat from one place to another.

magma: Melted rock within the earth.

mantle: The section of the earth between its crust and its core.

non-renewable: When something cannot be replaced by natural environmental cycles as quickly as we are using it.

permeable: Able to allow liquids or gasses to flow through.

radioactive: The state of giving off energy particles.

renewable: When something can be replaced by natural cycles in nature or the environment.

reservoir: A place where something, especially a liquid, is collected.

tectonic plate: A piece or section of the earth's crust.

turbine: A type of wheel that spins from a moving force, such as steam or running water.

EXPLORE FURTHER

Do Animals Use Geothermal Energy?

Investigate ways animals and plants use geothermal power. In your investigation, think about where and how these animals build their homes. Consider animals that live in deserts, cold climates, and near active geologic features, such as hot springs. How might they use the earth's heat? How might plants use the earth's heat? Write a short essay describing what you discovered.

Geothermal Map

Use a computer with access to the Internet and look up the location of geothermal power plants. Next, plot the power plants on a map of the world. Find a map of the earth's tectonic plates. Compare one map to the other. Are most geothermal plants located on the edges between plates? Did you find any that weren't? Are they near a hot spot or other geologically active feature? If you were an engineer, where might you explore for other sources of geothermal power?

A Geothermal Future?

Imagine you are the U.S. president. What would you tell the country's citizens about geothermal energy? What policies might you and Congress put in place to encourage people to use geothermal power? Is there a way you might use tax dollars to improve geothermal technology? What arguments would you use to convince people that geothermal research is worth the money and risk? What arguments would you use to dispute geothermal energy? Come up with a three-minute presentation to explain your position on geothermal energy.

Use a computer to do research so you can compare geothermal power plant locations with the earth's tectonic plates. What did you discover?

SELECTED BIBLIOGRAPHY

Blodgett, Leslie. "Oil and Gas Coproduction Expands Geothermal Power Possibilities." *Renewable Energy World.com*, July 10, 2010. Web. Accessed September 24, 2012.

Craddock, David. *Renewable Energy Made Easy,* Ocala, FL: Atlantic Publishing Group Inc., 2008. Print.

Lloyd, Donal Blaise. *The Smart Guide to Geothermal,* Masonville, CO: PixyJack Press, 2011. Print.

Tabak, John. *Solar and Geothermal Energy,* New York: Facts on File, 2009. Print.

FURTHER INFORMATION

Books

Mooney, Carla. *Geothermal Power*. San Diego: ReferencePoint Press, 2011.

Morris, Neil. *Geothermal Power*. Mankato, MN: Smart Apple Media, 2010.

Orr, Tamra. *Geothermal Energy*. Ann Arbor: Cherry Lake Publishing, 2008.

Reynoldson, Fiona. *Understanding Geothermal Energy and Bioenergy*. New York: Gareth Stevens Publishers, 2011.

Websites

http://geothermal.marin.org/index.html
This website features a map, slideshow, and more—all designed to teach you more about geothermal energy.

http://geo-energy.org/basics.aspx
What is geothermal energy? How does it work? Get the answers to these questions and more at the Geothermal Energy Association's website.

http://www.eia.gov/kids/energy.cfm?page=geothermal_home-basics
Learn more about how we find and use geothermal energy, as well as how geothermal power affects the environment.

INDEX